Marion Deuchars

DRAW This!

ART ACTIVITIES to unlock the IMAGINATION

LAURENCE KING

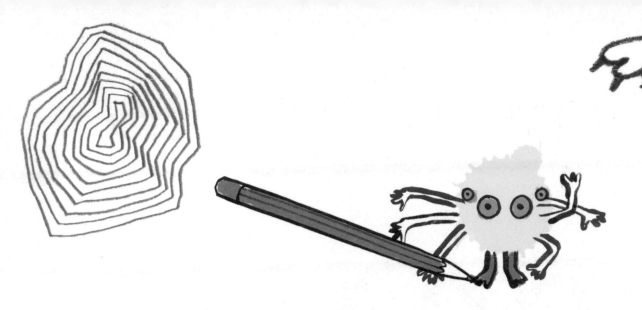

YOU CAN DRAW ANYTHING!

IN THIS BOOK, YOU'RE IN CHARGE! YOU CAN DRAW WHOEVER OR WHATEVER YOU WANT AND ALL OF YOUR PICTURES WILL TELL THEIR OWN AMAZING STORIES. SO LET YOUR MIND RUN FREE AND GET DRAWING!

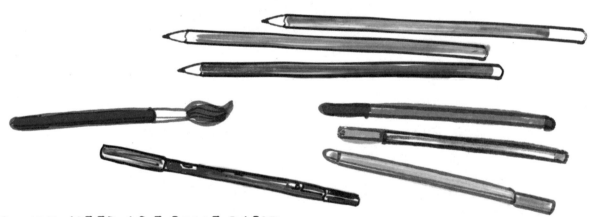

ALL YOU NEED ARE SOME BASIC DRAWING MATERIALS: PENS, PENCILS, COLORED PENCILS, PAINTS, AND BRUSHES.

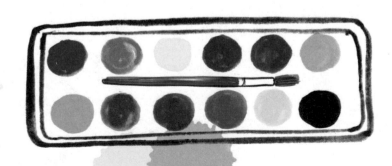

This book belongs to:

WHAT'S THAT HIDING BEHIND THE CLOUDS?
DRAW WHAT YOU IMAGINE IT OR THEY COULD BE . . .

HOW TO DRAW A RABBIT

1.

2.

3.

4.

5.

6.

DRAW YOUR OWN RABBIT. MAYBE YOUR RABBIT WANTS
TO BE A SUPERHERO. CAN YOU GIVE IT A SUPERHERO
OUTFIT AND NAME?

ONCE THERE WERE TWO FUNNY CREATURES THAT LIVED IN TWO STRANGE CASTLES. DRAW THEM HERE . . .

CAN YOU TURN THE FRUIT AND VEGETABLES INTO CHARACTERS?

NOW DRAW YOUR OWN IMAGINARY FRUIT.
WHAT IF YOU MIXED A BANANA WITH A PEA?

HOW TO DRAW AN OWL
*FOR THE WHITE HEART SHAPE, USE AN ACRYLIC
PEN OR CUT OUT WHITE PAPER.

1.

2.

3.

4.

5.

6.

A FAMILY OF OWLS ARE MOVING TO A FOREST IN THE DARK, DARK NIGHT. DRAW THEM IN THEIR NEW HOME. DO ANY OTHER CREATURES LIVE IN THE FOREST?

CAN YOU DRAW THE BODIES THAT BELONG TO THESE LEGS?
THEY MAY NOT BE FROM THIS PLANET!

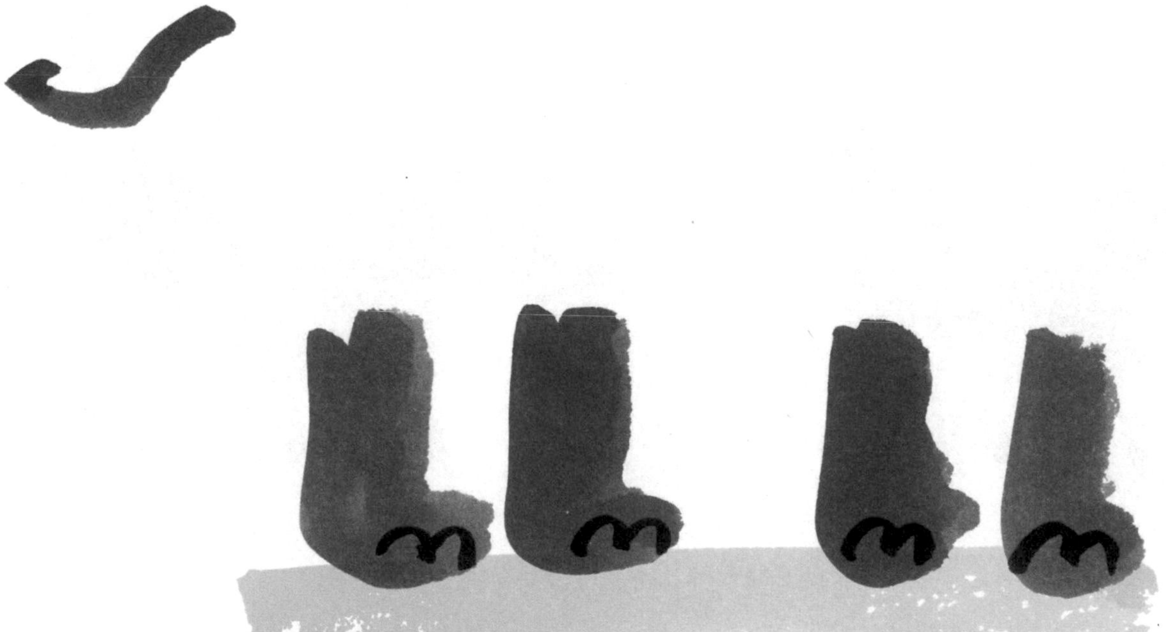

HOW TO DRAW A FOX

1.

2.

3.

4.

5.

6.

FOX IS HAVING A BIG PARTY WITH HIS FRIENDS.
WHAT ARE THEY CELEBRATING?

DRAW A BATTLE OF ALIENS OR MONSTERS IN THE MOUNTAINS.

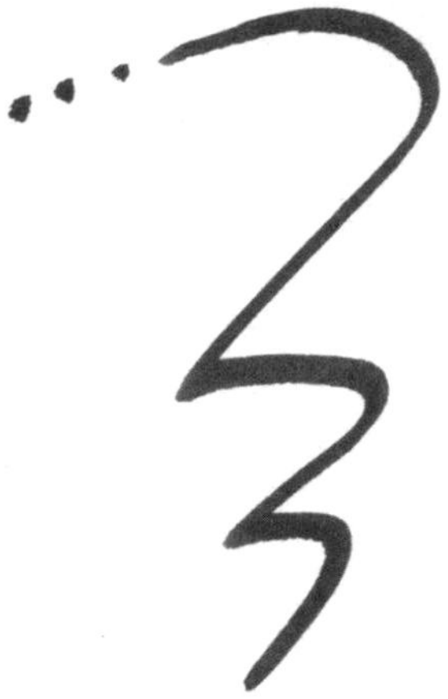

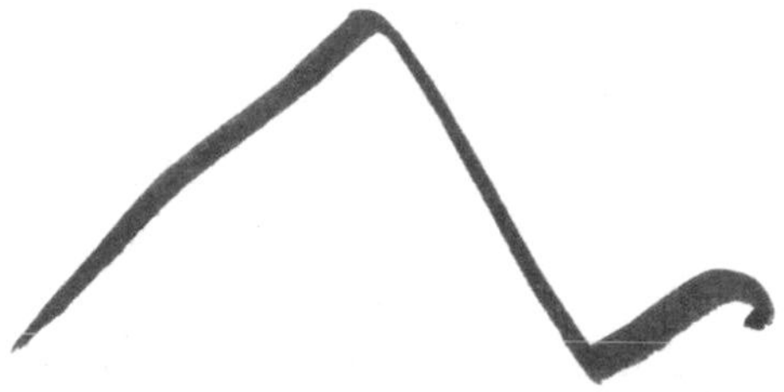

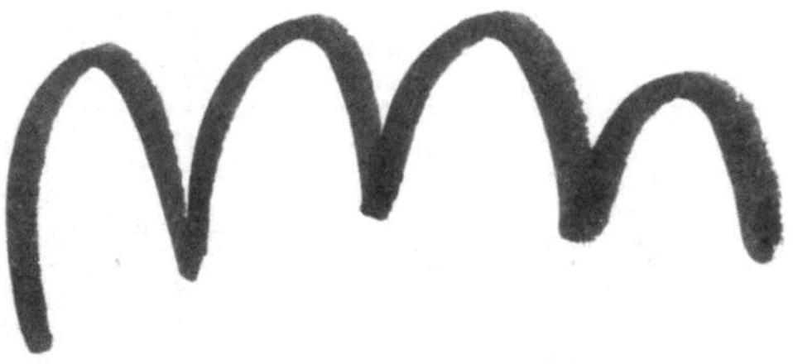

FINISH THIS DRAWING. IT DOESN'T HAVE TO BE A FISH!
IT COULD BE A MERMAID, A DOGFISH, OR A MADE-UP CREATURE!

DRAW A SECRET ISLAND THAT HAS ONLY JUST BEEN DISCOVERED.
WHO LIVES THERE? WHAT DO THE PLANTS AND BUILDINGS LOOK LIKE?

MAYBE THERE ARE THINGS IN THE WATER TOO!

THIS IS NO ORDINARY EGG. DRAW WHAT YOU THINK
IS INSIDE IT. (IT'S STILL A BABY THING)

NOW DRAW IT ONCE IT HAS GROWN UP.

WHAT IS THE BIRD CARRYING IN ITS FEET? IT'S BIG AND STRANGE!
DRAW IT . . .

HOW TO DRAW A SLEEPING DOG

1.

2.

3.

4.

5.

6.

NOW DRAW THE SLEEPING DOG IN THEIR VERY COMFORTABLE BED OR HOUSE.

HOW TO DRAW A BUTTERFLY

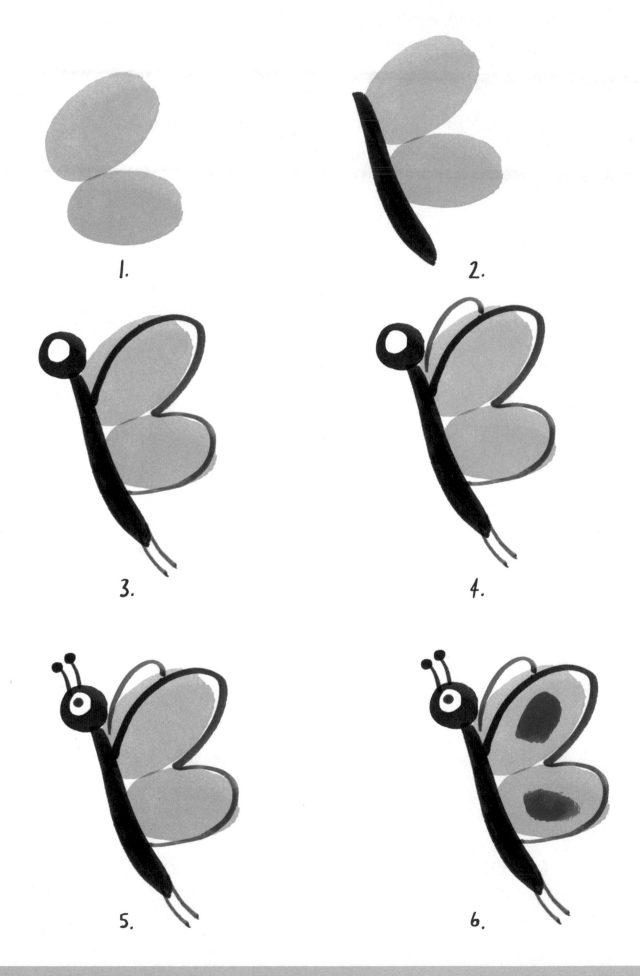

1.

2.

3.

4.

5.

6.

NOW DRAW A GARDEN FULL OF FLOWERS AND
BUTTERFLIES OF ALL DIFFERENT COLORS.

THESE TWO HOUSES ARE VERY DIFFERENT. CAN YOU DRAW THEM?

SOMETHING VERY UNUSUAL IS CLIMBING INTO THE BOX.

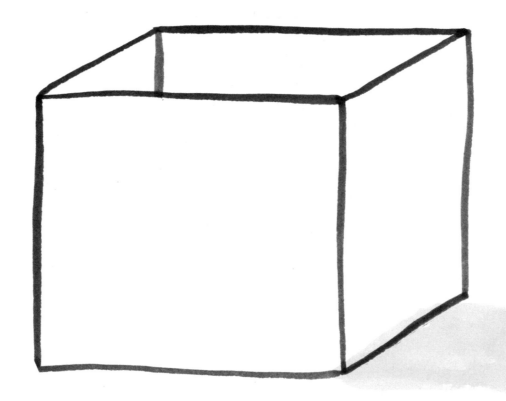

THE BOX HAS SOME STRANGE PATTERNS ON IT TOO!

SOMETHING VERY COLORFUL AND BIG
IS CLIMBING OUT OF THE BOX.

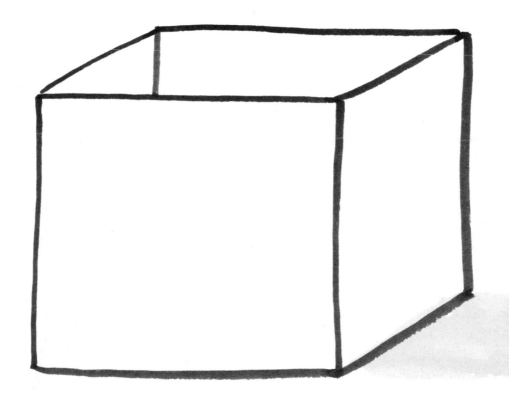

DRAW WHO IS CARRYING THESE STRANGE OBJECTS.
CAN YOU MAKE SOME OF YOUR OWN SHAPES TOO?

THESE ANTS ARE CARRYING SOME VERY LARGE THINGS ON
THEIR BACKS. CAN YOU DRAW THEM?

REMEMBER ANTS ARE VERY SMALL!

DRAW SOMETHING LARGE BEING CHASED BY SOMETHING
SMALL UP THIS HILL.

WHOSE TAILS ARE THESE? DRAW THE WILD AND WONDERFUL
CREATURES YOU THINK THEY BELONG TO.

HOW TO DRAW A BEAR

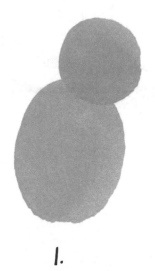

1.

2.

3.

4.

5.

6.

NOW DRAW A GROUP OF BEARS.
WHAT ARE THEY LOOKING AT IN THE POND?

FILL THIS PAGE WITH A FOREST FULL OF UNUSUAL AND
COLORFUL TREES AND PLANTS.

WHO LIVES IN THIS FOREST? ARE SOME OF THEM HIDING?
DRAW THEM TOO.

HOW TO DRAW A CAT

1.

2.

3.

4.

5.

6.

DRAW THE CAT ENJOYING ITS FAVORITE DINNER.
WHERE IS IT? IS IT AT HOME, OUTDOORS OR MAYBE EVEN
IN ITS FAVORITE CAFE?

IMAGINE THERE REALLY WAS LIFE ON MARS.
WHAT DO YOU THINK IT COULD LOOK LIKE?

YOU'VE DISCOVERED A WILD NEW WORLD IN SPACE.
WHAT COLOR IS IT? IS IT COVERED IN ICE OR
DOES IT MAYBE HAVE METALLIC CLOUDS?

HOW TO DRAW AN ELEPHANT

1.

2.

3.

4.

5.

6.

NOW DRAW SOME ELEPHANTS BALANCING ON TOP OF EACH OTHER TO MAKE A TOWER. MAYBE THEY'RE TRYING TO REACH SOMETHING VERY HIGH UP.

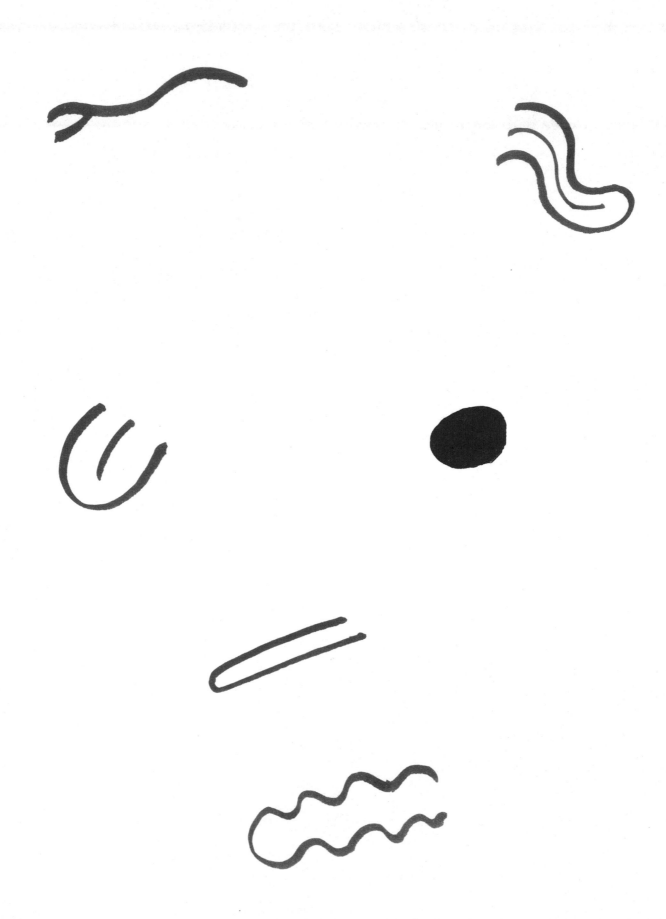

HOW TO DRAW A TIGER

1.

2.

*FOR THE WHITE SHAPES, USE AN ACRYLIC
PEN OR CUT OUT WHITE PAPER.

3.

4.

5.

6.

DRAW A TIGER HAVING A PICNIC WITH HIS FRIENDS.

DRAW A TIGER HAVING A PICNIC WITH HIS FRIENDS.

DRAW THE WILD AND CRAZY HEADS THAT THESE EYES BELONG TO.

LAURENCE KING

First published in Great Britain in 2023 by

Laurence King Publishing

10 9 8 7 6 5 4 3 2 1

Copyright © Marion Deuchars 2023

Marion Deuchars has asserted her right under the copyright, designs,
and patents act 1988, to be identified as the author of this work.

A CIP catalog record for this book
is available from the British Library.

ISBN 978-1-5102-3020-0

Printed and bound in China

Laurence King Publishing
An imprint of
Hachette Children's Group
Part of Hodder and Stoughton
Carmelite House
50 Victoria Embankment
London EC4Y 0DZ

An Hachette UK Company
www.hachette.co.uk
www.hachettechildrens.co.uk

www.laurenceking.com